FLORA OF TROPICAL EAST AFRICA

DILLENIACEAE

G. Ll. Lucas

Trees, shrubs, climbers, more rarely perennial herbs. Leaves alternate (opposite in *Hibbertia* from Madagascar), simple, rarely deeply lobed or pinnatifid (not in Africa). Stipules absent or more rarely adnate to the petiole and wing-like. Flowers bisexual, or unisexual, regular, hypogynous, borne in terminal or axillary cymes or racemes, sometimes solitary. Sepals 3–5(–20), free, imbricate, persistent. Petals (2–)5(–6), free, imbricate, often relatively large and showy. Stamens (3–)∞, free or partially united at the base into bundles, often persistent; anthers 2-thecous, dehiscence by means of a longitudinal slit or apical pore. Carpels (1–)3–5(–20), free or partially united along the central axis; ovules 1 or more, anatropous, erect from the base or inner suture; styles free; stigma simple. Fruit a follicle or berry-like. Seeds usually bearing a laciniate aril and copious fleshy endosperm.

A mainly tropical family of about 10 genera, with only one genus, *Tetracera* L., in Africa.

TETRACERA

L., Sp. Pl.: 533 (1753) & Gen. Pl., ed. 5: 237 (1754); Gilg in E.J. 33: 194 (1902); Gilg & Werderm. in E. & P. Pf., ed. 2, 21: 16 (1925)

Shrubs or climbers, rarely herbs or trees. Leaves petiolate; petiole often slightly grooved above; blade entire to denticulate, glabrous, tomentose or scabrid. Inflorescences few- to many-flowered terminal or axillary (usually from upper leaves) panicles. Flowers regular, ♀. Sepals 4–6(–15), imbricate, persistent, often reflexed in flower. Petals (2–)3–5(–6), usually white and showy, caducous, often emarginate. Stamens ∞, free; filaments thin at base, expanding towards apex; anther-thecae small and separated by the expanded connective, parallel, but more usually divergent towards their base, longitudinally dehiscent. Carpels 3–5, free; ovules usually numerous, more rarely reduced to 2, attached to the ventral suture; styles free, short; stigmas simple, only slightly differentiated from style. Fruit capsular, coriaceous, opening by the ventral and often the dorsal suture. Seeds 1–5, glossy, black in Africa, with a fimbriate or laciniate yellowish, reddish or purplish aril.

A pantropical genus of about 40 species, including 8 African, of which 5 are represented in East Africa.

Climber or scandent shrub:
  Leaves scabrid; carpels glabrous .   .   .   .   5. *T. potatoria*
  Leaves practically glabrous or sparsely hairy; carpels
    sparsely pubescent .   .   .   .   .   4. *T. litoralis*
Small tree, shrub or subshrub:
  Carpels glabrous .   .   .   .   .   .   2. *T. masuiana*
  Carpels pilose:
    Leaves bullate above; sepals subglabrous inside . 1. *T. boiviniana*
    Leaves not bullate above; sepals ± pubescent within 3. *T. bussei*

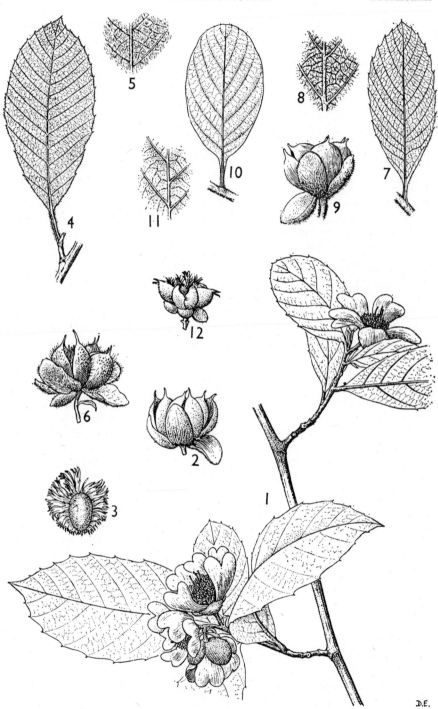

FIG. 1. *TETRACERA LITORALIS*—**1**, habit, × ⅔; **2**, fruit, × 1; **3**, arillate seed, × 2. *T. BOIVINIANA* —**4**, leaf, upper surface, × ⅔; **5**, leaf, lower surface, × 2; **6**, fruit, × 1. *T. MASUIANA*—**7**, leaf, upper surface, × ⅔; **8**, leaf, lower surface, × 2; **9**, fruit, × 1. *T. POTATORIA*—**10**, leaf, upper surface, × ⅔; **11**, leaf, lower surface, × 2; **12**, fruit, × 1. 1 from *Tweedie* 2383; 2, 3, from *Wallace* 703; 4, 5, from *Lucas* 244; 6, from *Tanner* 2559; 7, 8, from *Richards* 7413A; 9, from *Bally* 7531; 10, 11, from *Dawkins* 529; 12, from *Drummond & Hemsley* 4611.

1. **T. boiviniana** *Baill.*, Adansonia 7: 300, t. 7 (1867); Oliv., F.T.A. 1: 13 (1868), pro parte, excl. specim. *Welwitsch;* P.O.A. B: 330 & P.O.A. C: 272 (1895); V.E. 3(2): 476, fig. 220 (1921); Gilg & Werderm. in E. & P. Pf., ed. 2, 21: 17, fig. 7 (1925); T.T.C.L.: 182 (1949); Wild in F.Z. 1: 103, t. 7/B (1960); K.T.S.: 171 (1961). Type: Kenya, Mombasa, *Boivin* (P, holo.!)*

Shrubs or small trees up to 6(–7) m. high. Young branches villous, mature ones with a thin papery bark which splits and flakes off longitudinally. Leaf-blade elliptic to obovate, up to 13 cm. long, 6 cm. wide, acute to obtuse, broadly cuneate or sometimes ± rounded at base, slightly serrate, becoming more prominently toothed towards apex (each tooth with a small mucro), bullate, bicolorous; upper surface dark grey-green, sparsely pubescent, ± scabrid, the (9–)12–15(–18) secondary veins ± parallel and ending in a marginal mucro, tertiary veins reticulate, deeply impressed; lower surface pale grey-green, densely tomentose with the midrib and secondary veins prominent; petiole up to 1·5 cm. long, ± tomentose. Inflorescence cymose, terminal or axillary in upper branches; flowers sweet-scented, white or pinkish; peduncle and pedicels pilose. Sepals 4 or 5, broadly obovate to suborbicular, 10–12 mm. long, 8–10 mm. wide, silky pilose outside, glabrescent to slightly pilose within, persistent. Petals 4–5, broadly obovate, up to 2 cm. long, 1·5 cm. wide, usually emarginate, glabrous, caducous. Stamen-filaments 3–5 mm. long; anther-thecae divergent at their bases. Carpels 3–4–5, obovoid, densely to sparsely hirsute; ovules 1–5 per carpel; placentation parietal; styles free, filamentous, slightly thickened at the apex, up to 5(–6) mm. long. Fruiting carpels obovoid, slightly keeled along line at suture, up to 16 mm. long, 13 mm. across, woody, red when ripe, bearing remains of persistent style, ± sparsely hirsute. Seeds up to 5, black, glossy, reticulately patterned; aril fimbriate, up to 1 cm. long, creamy yellow. Fig. 1/4–6.

K<small>ENYA</small>. Kwale District: Diani, June 1934, *Napier* 6258! & Shimba Forest, 26 Apr. 1962, *Lucas, Jeffrey & Kirrika* 244! & Shimba Hill, 14 Jan. 1964, *Verdcourt* 3913!
T<small>ANGANYIKA</small>. Lushoto District: Korogwe, May 1958, *Semsei* 2742!; Tanga District: Pongwe, 31 Jan. 1937, *Greenway* 4853!; Uzaramo District: 16 km. W. of Dar es Salaam, 30 Nov. 1955, *Milne-Redhead & Taylor* 7451!
D<small>ISTR</small>. **K**7; **T**3, 6, 8; Mozambique coastal regions
H<small>AB</small>. Coastal woodland and thicket; 50–350 m.

2. **T. masuiana** *De Wild. & Th. Dur.* in Ann. Mus. Congo, Bot., sér. 1, 1: 61, t. 31 (1899); V.E. 3(2): 476 (1921); C.F.A. 1: 8 (1937); Staner in B.J.B.B. 15: 297 (1939); F.P.S. 1: 155 (1950); Wild in F.Z. 1: 104, t. 7/A (1960); F.F.N.R.: 247 (1962). Type: Congo, Lubunda, *Dewèvre* 1025 (BR, holo.!)

Erect shrub up to 1 m. high. Young branches densely pilose, yellowish; older branches with bark flaking into longitudinal patches, red-brown in colour. Leaf-blade oblanceolate to narrowly obovate, 4–16 cm. long, 1·8–6·5(–9) cm. wide, normally ± 10 cm. long, 4 cm. wide, ± rounded to mucronate, cuneate, strongly ± regularly serrate with mucro or glands prominent on teeth, densely and softly white to pale yellow pubescent above, densely villous beneath with usually white hairs; 7–12 pairs of secondary veins ± impressed above, prominent beneath, ending in marginal mucro; petiole short, ± winged, densely pilose. Inflorescence a few-flowered terminal cyme; flowers white, sweet-scented; peduncle and pedicels densely pilose; bracts ± lanceolate, up to 0·9 mm. long. Sepals 4(–5), circular to

---

* There is no evidence of *T. boiviniana* growing on Zanzibar I., as suggested by Baillon and Oliver. One of the two sheets of the holotype at Paris clearly shows the locality as Mombasa.

broadly obovate, varying in size, with the innermost one up to 13 mm. long and 9 mm. wide, pilose outside and subglabrous to sparsely pubescent inside. Petals 4(–5), broadly obovate, 1·6–2·3 cm. long, up to 1·5 cm. broad, emarginate, slightly pubescent on base of outer side, glabrescent, imbricate, caducous. Stamen-filaments up to 7 mm. long; anther-thecae divergent at their base. Carpels 3, glabrous; ovules 8–10 per carpel; placentation parietal (ovules in 2 rows); styles free, ± filamentous. Fruiting carpels obovoid, slightly keeled along suture, up to 2·1 cm. long and 1·5 cm. wide, woody, reddish, glabrous, bearing remains of persistent style. Seeds black, glossy; aril fimbriate, up to 4·5 mm. long. Fig. 1/7–9, p. 2.

TANGANYIKA. Buha District: Kasulu, Dec. 1954, *Forcus Nkotagu* 11!; Tabora District: near Urambo, 6 Oct. 1949, *Bally* 7531!; Ufipa District: Kipili, 31 Jan. 1950, *Bullock* 2365!
DISTR. T4; Zambia, Angola, Congo, Cameroun and Sudan Republics
HAB. Woodland, wooded grassland and scrub; 1000–1500 m.

SYN. [*T. boiviniana* sensu Oliv., F.T.A. 1: 13 (1868) quoad specim. *Welwitsch* tantum, *non* Baill.]
    T. strigillosa Gilg in E.J. 33: 196 (1902); V.E. 3(2): 476 (1921); Broun & Massey, Fl. Sudan: 96 (1929). Types: Sudan Republic, Equatoria Province, *Schweinfurth* 3985 (B, syn.†, K, isosyn.!) & Bahr el Ghazal Province, *Schweinfurth* 4048 & 4279 & 1915 (all B, syn.†)
    T. masuiana De Wild. & Th. Dur. var. *sapinii* De Wild., Cie. Kasai: 354 (1910). Type: Congo Republic, Katanga, *Sapin* (BR, holo.!)

3. **T. bussei** *Gilg* in E.J. 33: 197 (1902); V.E. 3(2): 476 (1921); T.T.C.L.: 182 (1949). Type: Tanganyika, Songea District, *Busse* 1282a (B, holo.†, EA, iso.!)

An erect subshrub up to 40 cm. tall; young branches long-pilose; older bark longitudinally ridged, probably flaking off. Leaf-blade oblanceolate to obovate, 4·0–7·5 cm. long, 1·5–3·5 cm. wide, acute to mucronate, cuneate, serrate-dentate, strigose on both surfaces; secondary veins 6–12, prominent beneath, only slightly or not impressed above, ending in a small gland (mucro) at apex of marginal teeth; petiole indistinct, winged, densely pilose. Flowers few (± 2–3), terminal, white; peduncle long (up to 3·5 cm.), densely long-pilose; pedicel 1–2 cm. long, densely long-pilose; bract lanceolate or ± narrowly elliptic, up to 4 mm. long. Sepals 4, ovate to obovate, innermost pair smaller and less pubescent than outer pair, inner up to 12 mm. long and 8 mm. wide, outer 14 mm. long and 9 mm. wide, all shortly pubescent to pilose on both surfaces but slightly denser outside. Petals 4, obovate, 1·5–2·3 cm. long, 1·1–1·8 cm. wide, emarginate, imbricate, caducous. Stamen-filaments up to 6 mm. long; anther-thecae divergent at their base. Carpels (3–)4, densely hirsute; styles free, elongated. Fruiting carpels not known.

TANGANYIKA. Songea District: Lukimwa R. [probably ESE. of Songea], Feb. 1901, *Busse* 1282a!
DISTR. T8; not known elsewhere
HAB. Wooded grassland; 2500–3000 m.

4. **T. litoralis** *Gilg* in E.J. 33: 197 (1902); V.E. 3(2): 475 (1921); T.T.C.L.: 182 (1949); K.T.S.: 171 (1961). Type: Tanganyika, Rufiji District, W. side of Mafia I., *Busse* 422 (B, holo.†, EA, iso.!)

Scandent shrub or climber up to 5 m. long. Young branches reddish-brown, very sparsely pilose; older bark grey-brown, flaking in longitudinal patches. Leaf-blade obovate, 11(–15·5) cm. long, 4(–6·5) cm. wide, shortly acuminate to apiculate, broadly cuneate, crenate to serrate, concolorous, glabrous when mature, very sparsely pilose, mainly beneath and towards base in young leaves; secondary veins 7–13 pairs, slightly to not impressed above, prominent beneath and ± parallel; tertiary veins reticulate and slightly prominent beneath; petiole short, up to 2–3(–5) mm. long, glabrous to very sparsely pilose. Inflorescence a raceme terminal on new leafy shoots; flowers few, white, sweet-scented; peduncle 1–2 cm. long and pedicels up to 10 mm. long, sparsely pilose. Sepals 4–5, broadly obovate, up to 14 mm. long, 10 mm. wide, coriaceous, adpressed pubescent outside, glabrous to glabrescent inside, persistent in fruit. Petals 4–5, ± broadly obovate, 1·7–2·0 cm. long, 1·1–1·3 cm. wide, usually emarginate, glabrous, imbricate, caducous. Stamen-filaments up to 6 mm. long, thickened towards apex; anther-thecae divergent at their base. Carpels 3–4, densely hirsute; ovules few per carpel; placentation parietal; styles free, filamentous. Fruiting carpels obovoid, slightly keeled, up to 1·7 cm. long, 1·5 cm. wide, bearing remains of persistent style, slightly hirsute. Seeds up to 5 per carpel, black, shiny; aril fimbriate, up to 8 mm. long, yellowish. Fig. 1/1–3, p. 2.

KENYA. Kilifi District: Mida, May 1929, *R. M. Graham* in *F.D.* 2135! & Arabuko Forest near Gedi, June 1962, *Tweedie* 2383!
TANGANYIKA. Uzaramo District: Pugu Forest Reserve, 10 Mar. 1964, *Semsei* 3702!; Rufiji District: Mafia I., 3 Apr. 1933, *G. B. Wallace* 703!; Ulanga District: near Ifakara, May 1959, *Haerdi* 262/0!
ZANZIBAR. Pemba I., near Mwembeduka, 19 Dec. 1930, *Greenway* 2763!
DISTR. **K7**; **T6**; **P**; not known elsewhere
HAB. Lowland rain-forest and dry evergreen forest; 0–50(–500) m.

5. **T. potatoria** *G. Don*, Gen. Syst. 1: 69 (1831); V.E. 3(2): 477 (1921); Gilg & Werderm., E. & P. Pf., ed. 2, 21: 18 (1925); Staner in B.J.B.B. 15: 299 (1939); T.T.C.L.: 182 (1949); F.P.S. 1: 154 (1950); F.W.T.A., ed. 2, 1: 180 (1954). Type: Sierra Leone, *Afzelius* (BM, holo.!)

Scandent shrub or climber up to 5 m. long. Young branches greenish-brown rapidly turning red-brown, scabrid with appressed reddish-brown hairs; older bark flaking off in longitudinal strips. Leaf-blade oblong to broadly elliptic, 3·5–14·0 cm. long, 2·5–5·5 cm. wide, entire to slightly undulate, in W. Africa often dentate towards apex which is obtuse to very broadly acute, rounded at base, scabrid above and beneath; secondary veins 6–9 pairs, ± parallel, impressed above, prominent beneath; midrib and secondary veins sparsely pilose; petiole 9–18 mm. long, slightly scabrid, sometimes winged towards base. Inflorescence paniculate, many-flowered, terminal on first year leafy shoots; flowers white, sweet-scented; pedicels densely pilose, indumentum rusty brown; bracts lanceolate to ± ovate. Sepals 4–5, varying in size, innermost pair up to 5 mm. long, glabrous to ± pubescent outside, densely silky within, persistent in fruit. Petals 4–5, obovate, not emarginate, up to 1 cm. long, 0·6 cm. wide, ± glabrous, sometimes with small patch of hairs towards base, imbricate, caducous. Stamen-filaments up to 6 mm. long; anther-thecae divergent at their base; carpels 3–4, hirsute. Ovules few per carpel; placentation parietal; styles free, filamentous. Fruiting carpels obovoid, up to 8 mm. long, 6 mm. wide, striate, keeled, bearing remains of persistent styles. Seeds 1 per carpel, up to 5 mm. in diameter, black, glossy; aril fimbriate, sheathing and slightly longer than seed. Fig. 1/10–12, p. 2.

UGANDA. Masaka District: Kyebe–Katera road, 4 Oct. 1953, *Drummond & Hemsley* 4611!; Mengo District: Kipayo, May 1914, *Dummer* 808! & 20 km. on Kampala–Entebbe road, Jan. 1931, *Snowden* 1944!
TANGANYIKA. Bukoba District: 88 km. S. of Bukoba, 5 km. down Mubunda turn-off, Feb. 1959 (sterile), *Procter* 1138!; Mwanza District: presumably on or near Ukerewe I., *Conrads* in *E.A.H.* 13295!
DISTR. **U2, 4**; **T1**; widespread from Sudan and Congo Republics into West Africa
HAB. Fresh-water swamp forest, remnant patches of lowland rain-forest and thicket; 1000–1500 m.